Asa Gray

**Natural Science and Religion**

Two lectures delivered to the Theological school of Yale college

Asa Gray

**Natural Science and Religion**
*Two lectures delivered to the Theological school of Yale college*

ISBN/EAN: 9783337129828

Printed in Europe, USA, Canada, Australia, Japan

Cover: Foto ©berggeist007 / pixelio.de

More available books at **www.hansebooks.com**

# NATURAL SCIENCE AND RELIGION

## TWO LECTURES

DELIVERED TO THE

THEOLOGICAL SCHOOL OF YALE COLLEGE

By ASA GRAY

NEW YORK
CHARLES SCRIBNER'S SONS
743 AND 745 BROADWAY
1880

*Copyright*, 1880,
BY ASA GRAY.

CAMBRIDGE:
UNIVERSITY PRESS: JOHN WILSON & SON.

# NATURAL SCIENCE AND RELIGION

## LECTURE I. — SCIENTIFIC BELIEFS.

I AM invited to address you upon the relations of science to religion, — in reference, as I suppose, to those claims of natural science which have been thought to be antagonistic to supernatural religion, and to those assumptions connected with the Christian faith which scientific men in our day are disposed to question or to reject.

While listening weekly — I hope with edification — to the sermons which it is my privilege and duty to hear, it has now and then occurred to me that it might be well if an occasional discourse could be addressed from the pews to the pulpit. But, until your invitation reached me, I had no idea that I should ever be called upon to put this passing thought into practice. I am sufficiently convinced already that the members of a profession know their own calling better than any one else can know it; and in respect to the debatable land which lies along the borders of theology and natural science, and which has been harried by many a raid from both sides, I am not confident that I can be helpful in composing strifes or in the fixing of boundaries; nor that you will agree with me that some of the encounters were inevitable, and some of the alarm groundless. Indeed upon much that I may have to say, I expect rather the charitable judgment than the full assent of those whose approbation I could most wish to win.

But I take it for granted that you do not wish to hear an echo from the pulpit nor from the theological class-room. You ask a layman to speak from this desk because you would have a layman's thoughts, expressed from a layman's point of view; because you would know what a naturalist comes to think upon matters of common interest. And you would have him liberate his mind frankly, unconventionally, and with as little as may be of the technicalities of our several professions. Frankness is always commendable; but outspokenness upon delicate and unsettled problems, in the ground of which cherished convictions are rooted, ought to be tempered with consideration. Now I, as a layman, may claim a certain license in this regard; and any over-free handling of sensitive themes should compromise no one but myself.

As a student who has devoted an ordinary lifetime to one branch of natural history, in which he is supposed to have accumulated a fair amount of particular experience and to have gained a general acquaintance with scientific methods and aims, — as one, moreover, who has taken kindly to the new turn of biological study in these latter years, but is free from partisanship, — I am asked to confer with other and younger students, of another kind of science, in respect to the tendencies of certain recently developed doctrines, which in schools of theology are almost everywhere spoken against, but which are everywhere permeating the lay mind — whether for good or for evil — and are raising questions more or less perplexing to all of us.

But our younger and middle-aged men must not think that such perplexities and antagonisms have only recently begun. Some of them are very old; some are old questions transferred to new ground, in which they spring to rankness of growth, or sink their roots till they touch deeper issues than before, — issues of philosophy rather than of science, upon which the momentous question of theism or non-theism eventually turns. Some on the other hand are mere *survivals*, now troublesome only to those who are holding fast to theological positions which the advance of actual knowledge has rendered untenable, but which they do not well know how to abandon; yet which, in principle, have mostly been abandoned already.

To begin with trite examples. Among the questions which disquieted pious souls in my younger days, but which have ceased to disquiet any of us, are those respecting the age and gradual development of the earth and of the solar system, which came in with geology and modern astronomy. I remember the time when it was a mooted question whether geology and orthodox Christianity were compatible; and I suppose that when, in these quarters, the balance inclined to the affirmative, it was owing quite as much to Professor Silliman's transparent Christian character as to his scientific ability. One need not be an old man to know that Laplace was accounted an atheist because he developed the nebular hypothesis, and because of his remark that he had no need to postulate a Creator for the mathematical discussion of a physical theorem; for a venerable and most religious astronomer, still living, who adopted this hypothesis in his "Exposition of certain Harmonies of the Solar System," published only five years ago, thought it needful to add an appendix, asking the question, "Is the nebular hypothesis, in any form, essentially atheistical in its character?" He answered it in the negative, but with the *salvo*, that " this hypothesis, having to do with a strictly azoic period, enforces no connection with 'the development theory' of the beginning or of the progress of life."

The great antiquity of the habitable world and of existing races was the next question. It gave some anxiety fifty years ago; but is now, I suppose, generally acquiesced in, — in the sense that existing species of plants and animals have been in existence for many thousands of years; and, as to their associate, man, all agree that the length of his occupation is not at all measured by the generations of the biblical chronology, and are awaiting the result of an open discussion as to whether the earliest known traces of his presence are in quaternary or in the latest tertiary deposits.

As connected with this class of questions, many of us remember the time when schemes for reconciling Genesis with Geology had an importance in the churches, and among thoughtful people, which few if any would now assign to them; when it was thought necessary — for only necessity could justify it — to bring the details of the two into agreement by extraneous suppositions and forced constructions of language, such as would now offend our critical and sometimes our moral sense. The change of view which we have witnessed amounts to this. Our predecessors implicitly held that Holy Scripture must somehow truly teach such natural science as it had occasion to refer to, or at least could never contradict it; while the most that is now intelligently claimed is, that the teachings of the two, properly understood, are not incompatible. We may take it to be the accepted idea that the Mosaic books were not handed down to us for our instruction in scientific knowledge, and that it is our duty to ground our scientific beliefs upon observation and inference, unmixed with considerations of a different order. Then, when fundamental principles of the cosmogony in Genesis are found to coincide with established facts and probable inferences, the coincidence has its value ; and wherever the particulars are incongruous, the discrepancy does not distress us,

I may add, does not concern us. I trust that the veneration rightly due to the Old Testament is not impaired by the ascertaining that the Mosaic is not an original but a compiled cosmogony. Its glory is, that while its materials were the earlier property of the race, they were in this record purged of polytheism and Nature-worship, and impregnated with ideas which we suppose the world will never outgrow. For its fundamental note is, the declaration of one God, maker of heaven and earth, and of all things, visible and invisible, — a declaration which, if physical science is unable to establish, it is equally unable to overthrow.

But, leaving aside for the present all questions of this sort, I proceed with the proper subject of this discourse; namely, the further changes in scientific belief, which have occurred within my own recollection, even since the time when I first aspired to authorship, now forty-five years ago.

There will be no need to go much beyond the line of subjects which it has been my business to study, in order to bring before you, in a cursory review, not indeed all the disturbing topics of the time, but quite enough of them for our purpose. For the changes which we have to consider are all more or less connected with the evolutionary theories which are now uppermost in the popular mind. In this presentation, it is best to set them forth in their simplest or most general form, divested of all theological or philosophical considerations, which have been or may be attached to them. I should rather say, to some of them. For the foundations, or at least the buttresses, of the now prevalent doctrine of the derivative origin of species mainly rest upon researches independently made, without speculative bias, being the general contributions to biological science in this century; the results of which have been accepted as far as made out without apprehension or other than scientific controversy.

Upon no one of these particular points has there been a completer change of view than upon the distinctness of the animal and vegetable kingdoms. The former conviction that these two kingdoms were wholly different in structure, in function, and in kind of life, was not seriously disturbed by the difficulties which the naturalist encountered when he undertook to define them. It was always understood that plants and animals, though completely contrasted in their higher representatives, approached each other very closely in their lower and simpler forms. But they were believed not to blend. It was implicitly supposed that every living thing was distinctively plant or animal; that there were real and profound differences between the two, if only they could be seized; and that increased powers of investigation — microscopical and chemical — might be expected to discover them. This expectation has not been fulfilled. It is true that the ambiguities of a hundred years ago are settled now. The zoöphytes are all remanded to their proper places, though the animal kingdom at first claimed more than belonged to it. But other, more recondite and insurmountable, difficulties arose in their place. The best, I am disposed to say the settled, opinion now is, that there are multitudinous forms which are not sufficiently differentiated to be distinctively either plant or animal, while, as respects ordinary plants and animals, the difficulty of laying down a definition has become far greater than ever before. In short, the animal and vegetable lines, diverging widely above, join below in a loop. Naturalists may help classification, but do not alter these facts, when they sever this loop arbitrarily at what they deem the lowest point, or when they cut away the whole loop, and form of it a separate kingdom — the *Protista* of Hœckel. The only objection to the latter is that the definition of this *tertium quid* from plant on the one hand and animal on the other is equally impracticable. One difficulty is removed only to have two in its place. The fact is, that a new article has recently been added to the scientific creed, — the essential oneness of the two kingdoms of organic nature. I crave your patience while I enter somewhat into particulars.

Not many years ago it was taught that plants and animals were composed of different materials: plants, of a chemical substance of three elements, — carbon, hydrogen, and oxygen; animals of one of four elements, nitrogen being added to the other three. The plant substance, named *cellulose*, because it formed the cell-walls, was supposed to constitute the whole vegetable fabric. It was known that all plants produced nitrogenous matter in the form of a compound of four elements; but this was thought to be merely a contained product, in a structureless condition, and to be not so much essential to the plant's life as to that of the animals which the plants nourished. It was known to be structure-building material for animals: it was not known to be essential plant-structure also. But it was soon ascertained that this quaternary matter of the animal body was chemically the same in the plant, was elaborated there, and only appropriated by the animal. Next it was found that it was physiologically and structurally the same in the plant, that it was *the* living part of the plant, that which manifested the life and did the work in vegetable as well as in animal organisms. This substance, which is manifold in its forms and protean in its transformations, has, in its state of living matter, one physiological name which has become familiar, that of *protoplasm*. The statement that " protoplasm is the physical basis of life " must be accepted as true. As Professor Allman puts it, "wherever there is life, from its lowest to its highest manifestations, there is protoplasm ; wherever there is protoplasm, there too is life," or has been. The cellulose or solid material which composes the bulk of a tree or herb did not produce the protoplasm contained in its living parts, as was formerly supposed, but the protoplasm produced the cellulose: the semiliquid and mobile matter within produced the cell-walls which enclose it. The walls or solid parts are to the protoplasm what the shell is to the oyster. The contents not only preceded the protective investment, but can exist and prosper apart from it, as many a mollusk does, as many a simple plant does throughout the earlier and most active period of its life. Indeed this slimy matter lives before and apart from any thing which can be called a living being. A formless, apparently diffluent and structureless mass is seen to exhibit the essential phenomena of life, — to move, to feed, to grow, to multiply. We have spoken of beings so low in the scale that the individuals throughout their whole existence are not sufficiently specialized to be distinctively plant or animal: yet these are definite in form and fixed in phase, are individual beings, though we may not determine to which kingdom they belong. But there is life in simpler shape,

"If shape it might be called that shape has none,
Distinguishable in member, joint, or limb,"

there is vital activity in that which has not attained even the semblance of individuality. Little lumps of protoplasm are these, with outline in a state of perpetual change, divisible into two or three or more, or two or three combining into one mass, either way without hindering or altering their manifestations. This living matter — of which *Bathybius*, if there be a *Bathybius*, or if it be any thing more than protoplasm of sponges, is one example — is said to have nothing more than molecular structure. It would be safer to say that the microscope has as yet revealed no organic structure.

The natural history of protoplasm has recently been well expounded by Professor Allman, late President of the British Association, a most judicious naturalist, of conservative tendency; and his address, which you have read or should read, saves me from further details, and enables me to proceed to other evidences of the substantial oneness of the two kingdoms of organic nature.

Cellulose makes up the bulk of a vegetable, and was thought to be its true element. But it is now known to be not even peculiar to it: it enters largely into the fabric of certain animals, not of the very lowest grade. Starch was equally regarded as a purely and characteristically vegetable production; and its presence, in ambiguous cases, has been taken as a test. But it follows the example of cellulose. Being a prepared material from which cellulose in the plant is made by a molecular change, we are not now surprised to learn that starch-grains of animal origin have been found. We cannot conceive any thing more characteristic of a vegetable than chlorophyll, the green of herbage; for in it the special work of the plant is done, — namely, the transformation of mineral matter into organic, under the light of the sun, this being the prerogative of vegetation. Now, not only does chlorophyll abound in many ambiguous microscopical organisms of fresh and salt water, which except for this would be taken for animals, but it has recently been detected in hydras and sea-anemones and planarias, which are as certainly animals as are oysters and clams. Nor can it be thought that they possess something merely resembling chlorophyll; for it performs the characteristic work of that peculiar substance, which, as I have said, is the characteristic work of vegetation. For the index and essential accompaniment of this work (*i.e.*, of the conversion of mineral into organic matter) is the evolution of oxygen gas from the decomposition of carbonic acid, water, &c., in which, if in any thing, vegetation consists. Now, the proof that what these animals possess is chlorophyll itself is demonstrated by their performance of the same function. They decompose carbonic acid and evolve oxygen gas, just as a green leaf does. Moreover, the chlorophyll has been extracted and identified by the spectroscopic test. Here, then, animals, undoubted animals, in addition to their own proper functions, take on the essential function of plants. There is no avoiding the conclusion that such animals are doing the duty of vegetables.

Although I make little account of it, I should not overlook a more empirical distinction between the two kingdoms which has also failed. The characteristic features of an animal were mouth and stomach. This is the normal correlation of an animal with its conditions. Having to feed on vegetable matter, or what has been vegetable matter, in solid as well as liquid form, a mouth opening into an internal cavity of some sort was the natural pattern, to which all animals were supposed to conform. But Nature, with all her fondness for patterns, will not be arbitrarily held to them. Entozoa feed like rhizophytes; and turbellarias and their relatives have no alimentary canal, — the food taken by what answers to mouth passing as directly into the general tissue as does the material which a parasitic root imbibes from its host, or an ordinary root from the soil.

While animals are thus overpassing the boundary in one direction, vegetables are making reprisals on the other. The rule is, that vegetables create organic matter, and animals consume it, producing none. But, while some animals produce some organic matter, some plants even among those of the highest grade feed wholly upon other plants, or even upon animals or their products. Like animals, some are herbivorous and some are carnivorous. That certain plants live parasitically upon other plants or upon animals, has long been too familiar to be remarkable. But that plants of the highest grade could capture or in some way take possession of small animals, extract and feed upon their juices, and appropriate these as nourishment, is essentially a recent wonder and a recently ascertained fact. Yet some of the facts which point to this conclusion are old enough; and the conclusion would probably have been reached years ago, except for the preconception that plants and animals were too distinct for interchange of functions. Now that we know they are not, and that the living structure in the two is fundamentally identical, what were formerly regarded as freaks of Nature are no longer mere wonderments, but parts of a system, and capable of being correlated with the rest by investigation. And investigation soon ascertained that this carnivorous attachment to the vegetable organism in *Dionæa* and *Drosera* was an organ for digesting as well as capturing animal food. Juices are imbibed by it directly, as in animals from the stomach; and nourishing solid parts are rendered soluble and assimilable by imbuing them with peptones or digestive ferments, analogous in composition and in action to the gastric juice of the higher animals.

Perhaps nothing in Nature can be more wonderful than all this; and nothing is more characteristic of the change which has come over scientific mind in our day than the manner in which such a discovery is received. The leading facts were well known a hundred years ago, and more. But, until recently, these phenomena were regarded as altogether anomalous; and such anomalies appear to have troubled nobody, except the framers of definitions. "*Lusus naturæ*" was a convenient phrase, and stood in the place of explanation, — as if the play of Nature was something apart from her work.

No one seems to have had any difficulty in believing that a few particular plants were endowed with faculties of which no other plants were sharers. The thoughtful naturalist of our day is in a different frame of mind. He expects to find that the extraordinary is only an extreme case of the ordinary; and he looks for instances leading up from the one to the other. I cannot tarry to explain how this expectation has directed observation and stimulated research in this particular field, and reached the result that these wonderful plants are distinguished only by higher degrees and more prominent manifestations of a power which is in some sort common to many or to all their brethren. We learn, even, that the germinating embryo of a grain of corn feeds upon and digests the solid maternal nourishment which surrounds it, and the humblest mould appropriates the organic matter which it attacks, by the aid of a peptone or inversive ferment, not different in nature and office from the gastric and other juices by aid of which we appropriate our daily meals.

It does appear also that the lowest organisms, which live a kind of scavenger life, by using over again dead or effete organic matter running to decay — but to some of which living juices come not amiss — have also the power, certain salts being given, of creating organic matter, and building up a fabric without sunlight and without chlorophyll. Here, then, is the simplest organic life, — in which, germs being given, *i. e.* first individuals of the sort supplied and placed in favorable surroundings, they increase and multiply into more, each to multiply again, and so on, in geometrical progression. From such lowly basis the two kingdoms may be conceived to rise, diverging as they ascend in separate lines, — the one developing close relations with sunlight and becoming the food-producing vegetable realm; the other, the food-consuming animal realm, which, dispensed from the labor of assimilation, and from the fixity of position which generally attends it, may rise to higher and freer manifestations of life. Such, at least, appear to be the relations of the two kingdoms to each other and to their common base; and such is the conception through which we may attain to an explanation of how it may be that members of each line possess so many characteristics of the other.

I have said, " germs being given," the forms increase and multiply. If asked, Whence the germs, and were they everywhere and always prerequisite? the scientific answer must be yes, so far as we know. Thus far, spontaneous generation, or abiogenesis, — the incoming of life apart from that which is living, — is not supported by any unequivocal evidence, though not a little may be said in its favor. However it may be in the future, here scientific belief stands mainly where it did forty-five years ago, only on a better-tried and firmer footing.

It remains to mention two supposed distinctions between vegetables and animals which were until recently prominent, but which are no longer criteria, even as between the higher forms of the two.

The first is the faculty of automatic movement, or — to take up the question only on the highest plane — the faculty of making movements in reference to ends. This is affirmed of animals, and is an undoubted faculty of all of them, but was long denied to plants, perhaps from a notion that such movements argued consciousness. But consciousness, in any legitimate sense of the term, pertains only to the higher animals. To show the breaking down of the distinction, it would suffice to contrast the rooted fixity and vegetative growth of very many lower animals with the free locomotion of most microscopic aquatic plants and of the germs of those not microscopic; but plants of the highest organization furnish obvious examples better suited to our purpose. Is there not an independent movement, in response to an external impression, and in reference to an end, when the two sides of the trap of *Dionæa* suddenly enclose an alighted fly, cross their fringe of marginal bristles over the only avenue of escape, remain quiescent in this position long enough to give a small fly full opportunity to crawl out, soon open if this happens, but after due interval shut down firmly upon one of greater size which cannot get out, then pour out digestive juices, and in due time re-absorb the whole? So, when the free end of a twining stem, or the whole length of a tendril, outreaches horizontally and makes circular sweeps, and secures thereby a support, to which it clings by coiling; when a tendril, having fixed its tip to a distant support, shortens itself by coiling, so bringing the next tendril nearer the support; when a free revolving tendril avoids winding up itself uselessly around the stem it belongs to, and in the only practicable way, namely, by changing from the horizontal to the vertical position until it passes by it, and then rapidly resumes its horizontal sweep, to result in reaching a distant support,— is it possible to think that these are not movements in reference to ends? You may say that all such movements are capable of explanation, or in time will be so; are the result of mechanism, and adjustments, and of common physical forces. No doubt; and this is equally true of every animal movement, not excepting those instigated by volition. "Still it moves," as the humbled Galileo said of the earth; and the idea that such movements are in reference to ends is not superseded by any yet devised explanation of the mechanism.

A remaining distinction between plants and animals was based on the relations they respectively sustain to the air we breathe. This has already been stated, and the exceptions noted; but the topic is resumed in order to bring to view the substantially different relations of the two kingdoms to physical force.

Plants give out oxygen gas, and thus purify the air for the respiration of animals. Animals, consuming this oxygen, breathe it back to the air in the form of carbonic acid. But the putting of this contrast is only another way of saying that plants produce organic matter and animals decompose it. The oxygen gas given out by sunlit foliage is just what is left over when carbonic acid is decomposed and the carbon enters into the composition of the vegetable matter then produced. This elaborated matter, more complex and unstable than the materials of which it was made, is the food of animals, is first appropriated, then decomposed by them, and in the decomposition the carbon is given back to the air recombined with the oxygen they inhale, the carbon again taking the oxygen which was separated from it by the plant. So respiration means decomposition; and this decomposition in the animal economy means organic material used up, work done, energy degraded. It means that the clock-weight which was wound up by the sun in the plant has run down. It means that, very much as the sun, shining on the earth and ocean, converts water into vapor and lifts it into the upper air, so the same luminary, shining upon the plant, there raises mineral matter to a higher and unstable state, in what we call organic products, — in both cases endowing the affected matter with a certain energy. The exalted matter in the one case falls at length as rain, perhaps directly into the ocean from which it was lifted, perhaps upon a mountain summit, where as snow or glacier-ice it may long remain poised and comparatively stationary. But sooner or later it falls into the rivulet and the river, and in its fall and flow it expends its endowment of energy, and *does work*, — turns wheels and spins or forges, if man so directs, — and, when it has reached stable equilibrium at the level of the ocean, it will have expended just the energy which was imparted to it in the raising. So the energy with which the sun endowed vegetable matter when it was raised to the organic state may be given up as heat when this matter is restored to its original condition by burning, or falls slowly back to the same condition in the process of natural decay; or the heat, like the falling water, may do mechanical work.

But also the organic material may be consumed in the plant itself. For the plant, like the animal, is a consumer. The only difference is that, whereas the animal is always and only a consumer and decomposer, the plant creates or composes likewise, and it produces vastly more than it consumes or decomposes. It decomposes only when it does mechanical work. But all its processes, all movements, all transformations, are work done at the expense of organized material and accumulated energy. Even the act of storing up solar force in the green herbage, or rather the changes connected with it, can only be done at a certain cost, though the cost is small in comparison with the gain. But every transference of material from one place or one state to another is done only by the decomposition and loss of some portion of it, — one part suffering that another may be changed and saved. When the germ feeds upon the maternal store in the seed, a considerable part is consumed in order to make the rest available; and the loss is made manifest, just as in the breathing of an animal or in the combustion of fuel, by the evolution of carbonic acid and of heat. The same thing in its measure occurs in the upbuilding of the fabric, the carrying of material high into the air, — into a tree-top, for instance; and in all the processes of flowering, and in storing up in the seed the richest products as an outfit for a new generation. Where visible movements take place, the quicker action is at equivalent cost. The sensitive tendril, which will coil promptly after the first brushing with my finger, will coil again only after an interval of rest, and upon the third or fourth excitation, or after a certain number of spontaneous revolutions, it falls exhausted.

But material endowed with energy in the plant is largely transferred as food to animals. It brings to them an energy which they may use, but did not originate.

Not many years ago, it was taken for granted that living things moved and had their being, and did their work, by strength of their own; that the power by which I strike a blow, or write on my paper, or move my lips in articulate speech, was somehow an original contribution to, rather than a directed use of, the common forces of physical nature. To all who have familiarized themselves with the facts of the case, the contrary is now substantially certain. The sun is the source of all motion and force manifested in life on the earth, and plants are the medium in which energy is exalted to the most serviceable state. The work done by living beings is at the expense of, and is measured by, the passage of so much matter from an unstable to a relatively stable equilibrium, by the coming together of molecules into closer and firmer positions, and by the attendant fall of so much energy from an exalted to a relatively degraded condition. So plants, animals, men, in all their doings, add nothing to and take nothing from the sum of physical force. Their prerogative is, each in its measure, to direct the application of physical force, and to direct it to *ends*.

The idea of ends involves that of individuality. The higher animals, and men among them, are complete individuals. We cannot make the idea of individuality any clearer than by adducing them as examples of it. In the lowest form of life, in those amorphous or indefinitely polymorphous "little lumps of protoplasm" which the biologists have made known to us, and even, perhaps, in a stratum or mass which takes the form of whatever bounds it, it is said that we may contemplate the phenomena of life in that which has no manifest individuality. What have we between these two extremes?

The first and simplest individuality is that of cells. Cell-doctrine, or the cellular composition of plants and animals, belongs wholly to the biological science of the last half-century, although the name is older, and some knowledge of the structure in plants is as old as the microscope. The homologizing of animals with plants in this regard began about forty years ago; and the doctrine of the individual life of cells is recent. Unfortunately the rather inappropriate name *cell* came into use before the structure was rightly understood, and may be misleading. It was given, naturally enough, to the walls circumscribing cavities in ordinary plant-tissue, before it was understood that the walls were not made and then filled, — before it was known that the contents are the living thing, and the wall an encasement or shell.

The substance of our recent knowledge is, that a plant is an aggregate of organic units, mostly of very small size; that these are to the herb or tree what the bricks and stones of this chapel are to the edifice. Only they "are living stones, fitly framed together" in organic growth, and their walls answer to the cement. Animals do not differ materially, except that the mortar is mostly of the same nature as the bricks, and there is a greater or at length complete fusion or confluence of the cells. The component material, the protoplasm, is essentially the same, as has already been stated.

But each aggregate, each ordinary plant or animal, begins as one cell, which is then the simple individual. This in growth and propagation divides itself into two, these two into four, these into sixteen, and so on, thus building up the structure, — a whole, of which the individual cells are component parts. The simplest plant begins in the same way with an initial cell, but this, instead of multiplying with cohesion into a structure, multiplies with separation into progeny. Other simple plants go on without separation to form a row of similar cells, which may casually fall apart into individuals or may remain connected; but in either case each has its own life, and does what the others do, so that the separation or the continued connection is a matter of indifference. But when, higher in the scale, structures are built up, what were individuals become parts or organs, or the thousandth or millionth part of an organ; then the life of the cells is their own no less, but their individuality blends in the common life of the aggregate. By increasing complexity of organization, with increasing subordination of parts and specialization of office, the highest plants and animals are composed. In them each unit or cell has its own life and its own nutrition, while also contributing to the common weal, — some by this function, some by that; but in the higher forms all are somehow controlled by a pervasive life and directed to common ends, — ends the more various, complex, and special, in proportion to the rank of the organism in the scale of being. So, too, the component cells become effete and die, while the aggregate life continues; and the continued structure, which is nothing but an aggregate, is somehow informed, animated, and operated by a common life of higher grade than that of any or all its components.

In numerous lower plants and animals we cannot definitely determine what are *organisms* and what are *organs*; in the herb or tree, and in the coral polypidom, organ, individual, colony are inextricably blended; in the higher animals subordination of parts to a whole is completely attained. All along the ascent that which controls and subordinates parts aggrandizes its manifestations. The lowest animals add very little to merely vegetative life, except greater sensitiveness to external impressions and more free and varied response; a step higher brings in a greater range of unconscious feeling; the higher brute animals have attained unto specific desires, affections, imagination, and the elements of simple thought; the highest, gifted with reflective reason, may make their own thoughts the subject of thought. So, our conception of individuality is from ourselves, conscious beings: it is carried down unqualified to the brute animals with which we are associated; it becomes vague and shadowy in plants, but still, somehow, the idea inheres throughout all organisms. The beginning of organization is individuation or tendency to individualize. The completed self is man.

Here let me interject a remark in correction of a common misapprehension as regards the nature of the simplicity of the lowest organisms. An animalcule and a unicellular plant, or the cellular components of common plants or animals, are simple indeed, comparatively. But the recent science which has brought out the close connection of the lower with the higher forms (and showed that through all "one increasing purpose runs") is also showing, in all the latest microscopic work, that the plant-cell and the animal-cell are really very complex structures, and the processes through which one cell becomes two, instead of being a simple bisection, prove to be very elaborate and wonderful. The further the investigation is carried under the modern microscope, the more complex and recondite does their structure and behavior appear to be. They seemed to be simple because they are small; but much of the simplicity vanishes upon intimate acquaintance. Wherefore, in view of recent discoveries of this sort, it is premature to conclude that the "little lumps of protoplasm" described by Hæckel are really destitute of organic structure. It is an illusion to fancy that the mystery of life is less in an amœba or a blood-corpuscle than in a man.

From individuals in themselves, let us pass to questions relating to their succession and kinds.

Plants and animals, each propagating their kind, produce lines of individuals, sustaining to each other the relation of parent and progeny. These lines are the *species* of the naturalist. Have the species come down from the beginning of life, unaltered or altered; or have there been successive creations?

Taking first the vegetable and animal kingdoms as a whole, it has long been well understood that ages upon ages have passed since the earth was stocked with living beings of numerous sorts. Kind after kind has appeared, flourished, and disappeared; and, in the long succession, species of progressively higher rank have come into existence, the forms more and more approximating those which now exist. There is good reason to believe that at more than one epoch the earth has been as fully stocked with species as it is now, and in equal diversity, except as to the highest types. What relation have these beings of the earlier and of the succeeding times sustained to each other and to the present inhabitants of the earth?

Half a century ago, when I began to read scientific books and journals, the commonly received doctrine was, that the earth had been completely depopulated and repopulated over and over, each time with a distinct population; and that the species which now, along with man, occupy the present surface of the earth, belong to an ultimate and independent creation, having an ideal but no genealogical connection with those that preceded. This view, as a rounded whole and in all its essential elements, has very recently disappeared from science. It died a royal death with Agassiz, who maintained it with all his great ability, as long as it was tenable. I am not aware that it now has any scientific upholder. It is certain that there has been no absolute severance of the present from the nearer past; for while some species have taken the place of other species, not a few have survived unchanged, or almost unchanged. And it is most probable that this holds throughout; for certain species appear to have bridged the intervals between successive epochs all along the line, surviving from one to another, and justifying the inference that species — however originated — have come in and gone out one by one, and that probably no universal catastrophe has ever blotted out life from the earth. Life seems to have gone on, through many and great vicissitudes, now with losses, now with renewals, and everywhere at length with change; but from first to last it has inhered in one system of nature, one vegetable and one animal kingdom, which themselves show indications of a common starting-point. As respects the vegetation, from which I should naturally draw illustrations, the nature and amount of the likeness between the existing flora and that of a preceding geological period has recently been summed up by Saporta in the statement that there is not a tree nor a shrub in Europe or North America which has not recognizable relatives in the fossil remains of the tertiary period. It is like visiting a country church-yard, where "The rude forefathers of the hamlet sleep," and spelling out, one by one, from mossed and broken gravestones, the names of most of the living inhabitants of the parish, — names differing it may be in orthography from those on the village signs; but, as of the people, so of the trees, it is beyond reasonable doubt that the later are descendants of the earlier.

The same holds true of animals; and the facts therefore point toward the conclusion that existing species in general are descended from tertiary ancestors. But if so they have mostly undergone change, and great change as we go farther back with the comparison. And there are many existing forms of which no fossil ancestor is known. What relation, if any, can these sustain to a by-gone flora or fauna? And with what reason do we predicate change of species in former times if they are not changeable now? This brings up the question of the fixity or variability of species.

Scientific opinion upon this point is not what it was thirty or forty years ago. Then it was generally, though not universally, believed that species are perfectly definite and stable; capable of variation, indeed, but only within circumscribed limits. Wherever it was difficult or impracticable to discriminate them, the difficulty was presumed to be, not in the things themselves, but in the imperfection of the naturalist's knowledge or acumen. There was the evidence of a good number of cases to show that species had not perceptibly altered in four or five thousand years, and of some having lasted for a vastly longer time. Hence it was an article of scientific faith that species on the whole were fixed now, and that probably they have come down essentially unaltered from the beginning, — a beginning which was wholly beyond the ken and scope of science, which is concerned with questions about how things go on, and has nothing to say as to how they absolutely began. The naturalists of that day might suppose — certainly many of them did suppose — that existing species may have come into being by other than direct supernatural origination, and, indeed, the foremost of them were well aware that the question of origin would have to be reargued at no distant day. But, so far, the various speculative attempts at explaining the mystery of the incoming of species had not been encouraging, and eminent naturalists deprecated all general theories of the sort, as at the best a waste of time. So the fixity and inscrutability of species — though silently doubted by some, and controverted by a few — was still the postulate of natural history; and more than one laborious naturalist has been known to declare that, if this fixity was not complete, natural history was not worth pursuing as a science.

There is now a different attitude toward this class of questions. First, the absoluteness of species is no longer taken for granted. That species have a stability, that every form reproduces after its kind, is obvious; but it is equally obvious that the similarity of its individuals is not complete. It had been assumed that the differences brought about by variation are always comparatively small, unessential, and limited. This is now partly doubted, and partly explained away.

In the first place, much of the popular idea of the distinctness of all species rests on a fallacy, which is obvious enough when once pointed out. In systematic works, every plant and animal must be referred to some species, every species is described by such and such marks, and in the books one species is as good as another. The absoluteness of species, being the postulate of the science, was taken for granted to begin with; and so all the forms which have been named and admitted into the systematic works as species, are thereby assumed to be completely distinct. All the doubts and uncertainties which may have embarrassed the naturalist when he proposed or admitted a particular species, the nice balancing of the probabilities and the hesitating character of the judgment, either do not appear at all in the record or are overlooked by all but the critical student. Whether the form under consideration should be regarded as a new species, or should be combined with others into a more generalized and variable species, is a question which a naturalist has to decide for the time being, often upon insufficient and always upon incomplete knowledge; and increasing knowledge and wider observation generally raise full as many doubts as they settle. This may not be so decidedly the case in zoölogy as in botany; but I incline to the opinion that there is no wide difference in this respect. The patient and plodding botanist spends much of his time in the endeavor to draw specific lines between the parts of a series the extremes of which are patently different, while the means seem to fill the interval. When he is addressed by the triumphant popular argument, "if one form and one species has been derived from another, show us the intermediate forms which prove it," he can only ejaculate his wish that this ideal vegetable kingdom was the one he had to deal with. Moreover when he shows the connecting links, he is told, " Then these are all varieties of one species; species are fixed, only with wider variation than was thought." And when he points to the wide difference between the extremes, as being greater than that between undoubted species, he is met with the rejoinder, " Then here are two or three or more species which undoubtedly have true distinctions, if only you would find them out." That is quite possible, but it is hardly possible that such fine differences are supernatural.

Some one when asked if he believed in ghosts, replied, No, he had seen too many of them. So I have been at the making and unmaking of far too many species to retain any overweening confidence in their definiteness and stability. I believe in them, certainly. I do not exactly agree that they " are shadows, not substantial things," but I believe that they have only a relative fixity and permanence.

You will ask if lack of capacity to interbreed is not a criterion of species. I must answer, No. As a matter of course individuals of widely diverse species cannot interbreed; those of related species not uncommonly do; but it is said that when they do interbreed the hybrid progeny is sterile. Commonly it is so, sometimes not. The rule is not sufficiently true to serve as a test, either in the vegetable or in the animal kingdom. The only practical use of the test is for the discrimination of the higher grade of varieties from species. Now in fact some varieties of the same species will hardly interbreed at all; while some species interbreed most freely, and produce fully fertile offspring. So the supposed criterion fails in the only cases in which it could be of service. All that can be said is, that whereas known varieties tend to interbreed with unimpaired and sometimes with increased fertility, distinct species of near resemblance tend not to interbreed at all; and between the two extremes there are all intermediate conditions. Here, as throughout organic nature, the extremes are far apart; the interval is filled with gradations.

What then is the substantial difference between varieties and species? Just here is the turning-point between the former view and the present. The former doctrine was that varieties come about in the course of nature, but species not; that varieties *became* what they are, but that species were *originally made* what they are. I suppose that, even before the day of Darwinism, most working naturalists were reaching the conviction that this distinction was untenable; that the same rule was applicable to both; and therefore that either varieties did not come in the course of nature, or that species did.

Perfectly apprehending the alternative and its consequences, Agassiz took the ground that varieties as well as species were primordial, or rather that the more marked forms called varieties by most naturalists were species, and therefore original creations. Rightly to understand his view, it must be taken along with his conception of species, as consisting from the very first of a multitude of individuals.

Other naturalists were looking to the opposite alternative, and were coming to the conclusion that species as well as varieties were natural developments. In botany, this conclusion was reached more than sixty years ago, through observation and experiment, by an English clergyman and naturalist, Herbert, afterward Dean of Manchester. He announced his conviction that "horticultural experiments have established, beyond the possibility of doubt, that botanical species are only a higher and more permanent class of varieties," and, consequently, that the genus is the progenitor of the species belonging to it. Others have reached the same conclusion by more speculative routes, and have deduced the theoretical consequences. But no marked impression was made until the hypothesis of natural selection, or the preservation of favored races in the struggle for life was promulgated, and supplied a scientific reason for the diversification of varieties into species. The principle brought to view is too obvious to have been wholly overlooked. It is interesting to notice that the earliest known anticipation of that principle which Darwin and Wallace developed almost simultaneously, was published sixty years ago, by Dr. Wells, the sagacious author of the theory of dew, who hit upon the idea of natural selection while resident in America. As abstracted by Mr. Darwin, who evidently takes delight in the discovery of these anticipations, the points which Dr. Wells made were substantially these : —

All animals vary more or less: agriculturists improve domesticated animals by selection. What is thus done by art is done with equal efficacy, though more slowly, by Nature, in the formation of varieties of mankind, fitted for the country which they inhabit, and in this way: Negroes and mulattoes enjoy immunity from certain tropical diseases, and white men a comparative immunity from those of cold climates. Under the variation common to all animals, some of the darker would be better adapted than the rest to bear the diseases of a warm country,— say, of tropical Africa. This race would consequently multiply, while the others would decrease, directly, because the prevalent diseases would be more fatal to them, and indirectly, by inability to contend with their more vigorous neighbors. Through the continued operation of the same causes, darker and darker races would prevail over the less dark, and in time would monopolize the region where they originated or into which they had advanced. Similarly would white races, to the exclusion of dark, be developed and prevail in cooler regions.

Now, this simple principle,— extended from races to species; from the present to geological ages; from man and domesticated animals to all animals and plants; from struggle with disease to struggle for food, for room, and against the diverse hardships which at times beset all living things, and which are intensified by the Malthusian law of the pressure of population on subsistence,— population tending to multiply in geometrical progression, while food can increase only in a much lower ratio, and room may not be increasable at all, so that out of multitudinous progeny only the few fittest to the special circumstances in each generation can possibly survive and propagate,— this is *Darwinism;* that is, Darwinism pure and simple, free from all speculative accretions.

Here, it may be remarked that natural selection by itself is not an hypothesis, nor even a theory. It is a truth, — a *catena* of facts and direct inferences from facts. As has been happily said, it is a truth of the same kind as that which we enunciate in saying that round stones will roll down a hill further than flat ones. There is no doubt that natural selection operates; the open question is, what do its operations amount to. The *hypothesis* based on this principle is, that the struggle for life and survival of only the fittest among individuals, all disposed to vary and no two exactly alike, will account for the diversification of the species and forms of vegetable and animal life, — will even account for the rise, in the course of countless ages, from simpler and lower to higher and more specialized living beings.

We need not here enter into any further explanation of this now familiar but not always well-understood hypothesis; nor need I here pronounce any judgment of my own upon it. No doubt it may account for much which has not received other scientific explanation; and Mr. Darwin is not the man to claim that it will account for every thing. But before we can judge at all of its capabilities, we need clearly to understand what is contained in the hypothesis; for what can be got out of it, in the way of explanation, depends upon what has gone into it. So certain discriminations should here be attended to.

Natural selection we understand to be a sort of personification or generalized expression for the processes and the results of the whole interplay of living things on the earth with their inorganic surroundings and with each other. The hypothesis asserts that these may account, not for the introduction of life, but for its diversification into the forms and kinds which we now behold. This, I suppose, is tantamount to asserting that the differences between one species and another now existing, and between these and their predecessors, has come to pass in the course of Nature; that is, without miracle. In these days, all agree that a scientific inquiry whether this may be so — that is, whether there are probable grounds for believing it (no thoughtful person expects to prove it) — is perfectly legitimate; and, so far as it becomes probable, I imagine that you might safely accept it. For the hypothesis, in its normal and simplest form, — when kept close to the facts, and free from extraneous assumptions — is merely this: —

Given the observed capacity for variation as an inexhaustible factor, assuming that what has varied is still prone to vary (and there are grounds for the assumption), and natural selection will — so to say — pick out for preservation the fittest forms for particular surroundings, lead on and diversify them, and, by continual elimination of the less fit, segregate the survivors into distinct species. This, you see, assumes, and does not account for, the impulse to variation, assumes that variation is an inherent and universal capacity, and is the efficient cause of all the diversity; while natural selection is the proximate cause of it. So it is the selection, not the creation of forms that is accounted for.

Darwinism does not so much explain why we have the actual forms, as it does why we have only these and not all intermediate forms, — in short, why we have *species*. There is of course a cause for the variation. Nobody supposes that any thing changes without a cause; and there is no reason for thinking that proximate causes of variation may not come to be known; but we hardly know the conditions, still less the causes now. The point I wish to make here is that natural selection — however you expand its meaning — cannot be invoked as the cause of that upon which it operates, *i. e.*, variation. Otherwise, if by natural selection is meant the totality of all the known and unknown causes of whatever comes to pass in organic nature, then the term is no longer an allowable personification, but a sheer abstraction, which meaning every thing, can explain nothing. It is like saying that whatever happens is the cause of whatever comes to pass.

We may conclude, therefore, that natural selection, in the sense of the originator of the term, and in the only congruous sense, stands for the influence of inorganic nature upon living things, along with the influence of these upon each other; and that what it purports to account for is the picking out, from the multitude of incipient variations, of the few which are to survive, and which thereby acquire distinctness.

There is a further assumption in the hypothesis which must not be overlooked; namely, that the variation of plants and animals, out of which so much comes, is indefinite or all-directioned and accidental. This, I would insist, is no fundamental part of the hypothesis of the derivation of species, and is clearly no part of the principle of natural selection. But it is an assumption which Mr. Darwin judges to be warranted by the facts, and in some of its elements it is unavoidable. Evidently if the innate tendency to vary upon which physical circumstances operate is indefinite, then the variations which the circumstances elicit, and which could not otherwise amount to any thing, must be accidental in the same sense as are the circumstances themselves. Out of this would immediately rise the question as to what can be the foundation and beginning of this long and wonderful chapter of accidents which has produced and maintained, not only for this time but through all biological periods, an ever-varying yet ever well-adapted cosmos.

But the facts, so far as I can judge, do not

support the assumption of every-sided and indifferent variation. Variation is somehow and somewhere introduced in the transit from parent to offspring. The actual variations displayed by the progeny of a particular plant or animal may differ much in grade, and tend in more than one direction, but in fact they do not appear to tend in many directions. It is generally agreed that the variation is from within, is an internal response to external impressions. All that we can possibly know of the nature of the inherent tendency to vary must be gathered from the facts of the response. And these, I judge, are not such as to require or support the assumption of a tendency to wholly vague and all-directioned variation.

Let us here correct a common impression that Darwinian evolution predicates actual or necessary variation of all existing species, and counts that the variation must be in some definite ratio to the time. That is not the idea, nor the fact. "Evolution is not a course of hap-hazard and incessant change, but a continuing re-adjustment, which may or may not, according to circumstances, involve considerable changes in a given time." Every form is in a relatively stable equilibrium, else it would not exist. Forms adjusted to their surroundings ought by the hypothesis to remain unchanged until the circumstances change. Only those of their variations could come to any thing which happened to be equally well adapted to the unchanged circumstances; and this may be what we have when two or more nearly related species inhabit similar stations in the same area.

From this point of view you see how wide of the mark are those who imagine that Darwinian evolution supposes that the organic world was in early times, or at any time, out of joint or in ill relations to the surroundings. On the contrary, it is of the very nature of natural selection, that, while inducing changes eventually immense, it should preserve throughout all time a condition of harmonious adaptation. Catastrophes must destroy; but gradual modification, under the long and silent struggle which never hastes and never rests, preserves while it renovates and diversifies the races.

I ought here to state that there are eminent naturalists (one of them of your own university) who accept the doctrine of evolution, but who think little of natural selection as a *modus operandi* in the diversification of species; and there are distinguished writers, not naturalists, who, from other points of view are ready to accept "the doctrine of the successive evolution from ancestral germs of higher and higher forms of life and mind,"* while they profess to have buried the principle of natural selection and with it the Malthusian theory of population in one common grave. These are evolutionists, in their way, because the probability of evolutionary theories springs from the very various lines of facts, otherwise inexplicable, which they harmonize and explain : — in geology, the previous existence of forms more and more like those now existing, and at length coalescing in them; in geography, the actual distribution of species and genera over the earth's surface; in systematic natural history, the reason why species and genera and orders are so variously related, are here connected by transitions and there separated by wide gaps; in morphology why the same functions may be assumed by different organs, or the same kind of organ may perform here one function and there another, or again exist as a vestige, of no service at all; in anatomy and biology, the transition from one element of structure to another, the gradual

* Bowen in " The North American Review," November, 1879.

specialization of organs, and the remarkable coincidence between the order of the development in the individual animal and that of the rise from low to high in the scale of being, and that of the successive appearance of the grades in time; finally in psychology, the gradations between beings endowed with rudimentary sensation and beings endowed with mind.

Here, where the "touch of Nature makes the whole world kin," we reach the sensitive point. Man, while on the one side a wholly exceptional being, is on the other an object of natural history,—a part of the animal kingdom. If you agree with Quatrefages that man is a kingdom by himself, you must agree with him that this kingdom is solely intellectual; that he is as certainly and completely an animal as he is certainly something more. We are sharers not only of animal but of vegetable life, sharers with the higher brute animals in common instincts and feelings and affections. It seems to me that there is a sort of meanness in the wish to ignore the tie. I fancy that human beings may be more humane when they realize that, as their dependent associates live a life in which man has a share, so they have rights which man is bound to respect.

Man, in short, is a partaker of the natural as well as of the spiritual. And the evolutionist may say with the apostle: "Howbeit that was not first which is spiritual, but that which is natural, and afterward that which is spiritual." Man, "formed of the dust of the ground," endowed with "the breath of life," "became a living soul." Is there any warrant for affirming that these processes were instantaneous?

As has just been intimated, the characteristic of that particular theory of evolution which is now in the ascendant is that, by taking advantage of "every creature's best" for bettering conditions, it has made strife work for good, throughout an immensely long line of adjustments and readjustments, in a series ascending as it advanced; that it supposes a process, not from discord to harmony, but from simpler to fuller and richer harmonies, conserving throughout the best adaptations to the then existing conditions. So while its advocates nowhere contemplate a state

"When Nature underneath a heap,
 Of jarring atoms lay,
 And could not heave her head,"

they may appropriate Dryden's closing lines, —

> "From harmony, from heavenly harmony,
>   This universal frame began, .
> From harmony to harmony
>   Through all the compass of the notes it ran,
>   The diapason closing full in man."

I have now indicated, at more than sufficient length for one discourse, some of the principal recent changes and present tendencies in scientific belief, especially in biology. Even the most advanced of the views here presented are held by very many scientific men, — some as established truths, some as probable opinions. There is a class, moreover, by whom all these scientific theories, and more, are held as ascertained facts, and as the basis of philosophical inferences which strike at the root of theistic beliefs.

It remains to consider what attitude thoughtful men and Christian believers should take respecting them, and how they stand related to beliefs of another order. That will be the topic of a following lecture.

## LECTURE II.—THE RELATIONS OF SCIENTIFIC TO RELIGIOUS BELIEF.

IN a preceding discourse I brought to your notice a series of changes in view and opinion which have taken place among scientific men within my own remembrance. I restricted the survey to the biological sciences (with merely a reference to the principle of the conservation of energy in its application to the organic world), and in these to the supposed facts and immediate inferences, to what may be called their natural-historical interpretation.

These new views are full of interest of a kind which you cannot expect a naturalist to undervalue. For they have greatly exalted his calling. In the days of Linnæus, who died only a hundred and two years ago, and throughout a long generation of his followers, species were looked upon as "simple curiosities of Nature," to be inventoried and described; and striking phenomena in plants and animals, as something to be wondered at, but not to be explained. With the advent of Morphology, the precursor and parent of Evolution, Natural History developed from a curious pursuit, training the observing powers, to that of a true science, engaging the reason in the search for causes. According to one definition, "Science is the labor of mind applied to Nature." In this sense, modern botany and zoölogy have certainly become scientific. They are at least attempting great labors. But in widely extending, as they now do, the operation of natural causes in the organic world, they make close connections between biology and physics, or what used to be called, and I think deserves to be called, natural philosophy. And the connection brings in, or brings up afresh, considerations which affect the ground of natural and revealed religion. Under this aspect, they properly excite your anxious attention.

I used throughout the phrase "scientific belief," as the one best suited to the occasion. The term is comprehensive and elastic, covering many degrees of conviction or assent, from moral certainty down to probable opinion. In this respect, scientific and theological beliefs are similar; as they also are in being mainly states of mind toward that which is incapable of demonstration, — either because, as in the case of ultimate beliefs (on which all science and knowledge are based) it is impossible to go beyond them, or else because the subject-matter is not positively known, and certainty is unattainable from the nature or the present conditions of the case. The proofs upon which both biological and theological investigations have to rely are largely probabilities, some of a higher, some of a lower order, and much that is accepted for the time is taken on trial or on *prima facie* evidence. Much also is or should be held under suspense of judgment, a state of· mind eminently favorable to accurate investigation. As to those who can forthwith assort the contents of their minds into two compartments, one for what they believe and the other for what they disbelieve, neither their belief nor their denial can be of much account. In all subjects of inquiry, those only are to be trusted who discriminate between inevitable beliefs, established convictions, probable opinions, and hypotheses on trial.

Now, our general inquiry in this lecture is, What should be the attitude, I will not say of theological students, but of thoughtful men, in respect to scientific beliefs, tendencies, and anticipations, such as we have been considering?

To a certain extent it may well be a waiting attitude. The strictly scientific matters must necessarily be left mainly to the experts, whose very various and independent investigations, pursued under every diversity of bias, must in time reach reasonably satisfactory conclusions. But the naturalists claim no monopoly in the consideration of the great problems which now interest us, in which indeed most of them decline to take any part. Perhaps theological students and divines might be asked to wait until views and hypotheses still ardently controverted among scientific investigators are brought nearer to a settlement. But the disposition to discount expected results, either for or against supernatural religion, has always prevailed. The theologians at least have never waited, and cannot be expected to wait; and while some of their contributions to the subject have been inconsiderate, others have been most valuable.

In any case, there is no call to wait on the ground that the disturbing views are only hypotheses. For, in the first place, we should have long to wait for demonstration one way or the other; and one crop of hypotheses is the fertile seed of another. Besides, hypothesis is the proper instrument for dealing with this class of questions; indeed, it is the essential precursor of every fruitful investigation in physical nature. You can seldom sound with the plummet while standing on the shore. To do this to any purpose, you must launch out on the sea, and brave some risks. Nearly all valuable results have been gained in this way. Newton's theory of gravitation was a typical hypothesis, and one which happened to be capable of early and sufficient verification. The undulatory theory of light was another. The nebular hypothesis, or portions of it, and the kinetic theory of gases, less verifiable, are accepted willingly because of the success with which they explain the facts. Evolution is a more complex, loose, and less provable hypothesis, or congeries of hypotheses, which can at most have only a relative, though perhaps continually increasing probability from its power of explaining a great variety of facts. Its strength appears on comparing it with the rival hypothesis — for such it is — of immediate creation, which neither explains nor pretends to explain any.

How the more exact physical sciences are becoming more reconditely hypothetical, especially in the imagination of entities of which there can be no possible proof beyond their serviceability in explaining phenomena, we must not stop to consider. Only this may be said, that the adage, "Where faith begins science ends" is now well nigh inverted. For faith, in a just sense of the word, assumes as prominent a place in science as in religion. It is indispensable to both.

Let it be noted, moreover, that the case we have to consider does not come before the tribunal of reason with antecedent presumptions all on one side, as theologians generally suppose. They say to the naturalists, not improperly, we will think about adopting your conclusions, contrary as they are to all our prepossessions, when they are thoroughly and irrevocably substantiated, and not till then. Your theory may prove true, but it seems vastly improbable. Here the naturalist is ready with a rejoinder: In this world of law you cannot expect us to adopt your assumption of specific creations by miraculous intervention with the course of Nature, not once for all at a beginning, but over and over in time. We will accept intervention only when and where you can convincingly establish it, and where we are unable to explain it away, as in the case of absolute beginning. If the naturalist starts with the presumption against him when he broaches the theory of the descent of later from preceding forms in the course of Nature, so no less does the theologian when in a world governed by law he asserts a break in the continuity of natural cause and effect.

But, indeed, you are not so much concerned to know whether evolutionary theories are actually well-founded or ill-founded, as you are to know whether if true, or if received as true, they would impair the foundations of religion. And, surely, if views of Nature which are incompatible with theism and with Christianity can be established, or can be made as tenable as the contrary, it is quite time that we knew it. If, on the other hand, all real facts and necessary inferences from them can be adjusted to our grounded religious convictions, as well as other ascertained facts have been adjusted, it may relieve many to be assured of it.

The best contribution that I can offer towards the settlement of these mooted questions may be the statement and explanation of my own attitude in this regard, and of the reasons which determine it.

I accept substantially, as facts, or as apparently well-grounded inferences, or as fairly probable opinions, — according to their nature and degree, — the principal series of changed views which I brought before you in the preceding lecture. I have no particular predilection for any of them; and I have no particular dread of any of the consequences which legitimately flow from them, beyond the general awe and sense of total insufficiency with which a mortal man contemplates the mysteries which shut him in on every side. I claim, moreover, not merely allowance, but the right to hold these opinions along with the doctrines of natural religion and the verities of the Christian faith. There are perplexities enough to bewilder our souls whenever and wherever we look for the causes and reasons of things; but I am unable to perceive that the idea of the evolution of one species from another, and of all from an initial form of life, adds any new perplexity to theism.

In unfolding my thoughts upon the subject, I wish to keep as close "to the solid ground of Nature" as I possibly can, even where the discourse must rise from the ground of science into the finer air of philosophy. Specially I must heed the injunction: "If thou hast any tidings, prithee, deliver them like a man of this world," and not trouble myself, nor you, with metaphysical refinements and distinctions which, however needful in their way and place, are unnecessary to our purpose. I take for granted, "like a man of this world," the objective reality and substantiality of what we see and deal with, though I am told it cannot be proved; and I assume, — although demonstration is impossible, — that what I and my fellow-men cannot help believing we ought to believe, or at least must rest content with. I suppose you will agree with me that it is not science, at least not natural science, which raises the most formidable difficulties to Christian theism, but philosophy, and that it is for philosophy to surmount them.

The question which science asks of all it meets is, What is the system and course of things, and how is this or that a part of it in the fixed sequence of cause and effect? Philosophy asks whence the system itself, and what *are* causes and effects. Theology is partly historical science, and partly philosophy. Now I, as a scientific man, might rest in the probability of evolution as a general inference from the facts or a good hypothesis, and relegate the questions you would ask to the philosophers and theologians. But I am not one of those who think that scientific men should not concern themselves with such matters; and having gone so far as to say that the evolution which I accept does not seem to me to add any new perplexity to theism, and well knowing that others are of a contrary opinion, I am bound to further explanation and argument.

But I have not the presumption to suppose that I can make any new contribution to this discussion; and what I may suggest must not be expected to cover the ground widely nor penetrate it deeply. I am sure that you will not look to me for the rehandling of insoluble problems and inevitable contradictions, into which the philosophical consideration of the relations of Nature and man to God ultimately lands us. Certainly they are not peculiar to evolution. So, in so far as we may fairly refer any of its perplexities to old antinomies, which can neither be reconciled nor evaded, the burden will be off our shoulders. It might suffice to show that evolution need raise no other nor greater religious or philosophical difficulties than the views which have already been accepted, and held to be not inimical to religion.

But, indeed, our universal concession that *Nature is*, and that it is a system of fixed laws and uniformities, under which every thing we see and know in the inorganic universe, and very much in the organic world, have come to be as they are, in unbroken sequence, implicitly gives away the principle of all ordinary objection to the evolution of living as well as of lifeless forms, of species as well as of individuals. It leaves the matter simply as one of fact and evidence. Indeed, mediate creation is just what the thoughtful and thorough observer of the ways of God in Nature would expect, and is what some of the most illustrious of the philosophic saints and fathers of the church have more or less believed in.

In saying that the doctrine of the evolution of species has taken its place among scientific beliefs, I do not mean that it is accepted by all living naturalists; for there are some who wholly reject it. Nor that it is held with equal conviction and in the same way by all who receive it; for some teach it dogmatically, along with assumptions, both scientific and philosophical, which are to us both unwarranted and unwelcome; more accept it, with various confidence, and in a tentative way, for its purely scientific uses, and without any obvious reference to its ultimate outcome; and some, looking to its probable prevalence, are adjusting their conditional belief in it to cherished beliefs of another order. One thing is clear, that the current is all running one way, and seems unlikely to run dry; and that evolutionary doctrines are profoundly affecting all natural science.

Here you remark that your objection is not so much to the idea of mediate creation as to the form it has assumed; that the mediate production of species *may* indeed be completely theistic. But that, whereas their immediate creation directly asserts Divine action, their incoming under Nature only implies it. To those who already believe in a Supreme Being the two views may religiously amount to the same thing. But, you continue, living beings were thought to afford a kind of demonstration of a supernatural creator. Science, in taking this away, leaves us only the assurance that if we bring the idea of God to Nature we may find Nature wholly compatible with that idea. Well, what is lost in directness may perhaps be gained in breadth and depth. It is certain that the whole progress of physical science tends, in respect to Divine action, to consider that mediate, general, and in a sense indirect, which had been thought to be immediate and special. Youth is ever taught by instances, manhood by laws.

You go on to say: The evolution of species now so commended to us by science, not long ago seemed as improbable to scientific as to ordinary minds. What assurance can we unscientific people have that science will not reverse its present judgments? · None, perhaps, except that, while many particular judgments have been reversed or altered, the general course of thought has run in one direction. And theologians, like naturalists, must be content with the best judgments they can form upon the present showing, and be ready to modify them upon better.

Finally, and to reach the present point, you pertinently commend to scientific men their own saying: "Science asks of every thing how it is a part of the system of Nature, of the chain of cause and effect." An hypothesis must give the how and why, and from its own resources, before it is worth attending to. A credible hypothesis should assign real and known causes, and ascertain their actual operation somewhere before assuming their operation everywhere. A complete hypothesis should assign not only real but sufficient causes for all the effects; and when it assumes them in invisible and intangible forms, such as molecules and molecular movements, it is bound to show that all the observed consequences flow from the assumption. Now to declare that species come through evolution, without either proving it by facts or clearly conceiving the mode and manner how, is only supporting a thesis which was until lately deemed scientifically improbable by hypotheses of a kind which have always been regarded as invalid.

Just here Darwinism comes in with a *modus operandi*, in which lies all its essential value. As the conception of the derivation of one form from another is the only distinctly-pointed alternative to specific supernatural creation, so the principle of natural selection, taken in its fullest sense, is the only one known to me which can be termed a real cause in the scientific sense of the term. Other modern hypotheses assign metaphysical, vague, or verbal causes, such as development, anticipation, laws of molecular constitution, without indicating what the special constitution is, — none of which have much advantage over the "*nisus formativus*" of earlier science.

I have no time to recapitulate what I briefly said of natural selection in a former lecture; nor to analyze the applications of the principle by Darwin, Wallace, and others to critical instances; nor to specify its limitations and apparent failures. The discussion or even the presentation of these would fill the hour, and divert me from my particular task. Instead of this, I will merely give my impression of the present state of the case as respects the points now before us.

You will remember the distinction which I pointed out between the *principle* of natural selection, which I take to be a true one, and the Darwinian *hypothesis* founded on it, which I take to be to a considerable extent probable. That is, I think that the influences and actions which the term "natural selection" stands for, give a sufficient scientific explanation of the way in which smaller differences among plants and animals may rise into greater, varieties into species. Given differences and an internal tendency to differ more, *i.e.*, given variation as an inexhaustible factor, and natural selection should suffice for the preservation and increase of the select few as a consequence of the destruction of the intermediate many. Surely there is nothing either improbable or irreligious in the idea that lines of individuals or races, once in existence, should be subject to the conditions of Nature, and that the fittest for particular conditions should thereby be preserved. As to variation, that really occurs as a fact, though we know not how; and, if we frame explanations of the mode and get conceptions of the causes of the variation of living things, still we probably shall never be able to carry our knowledge very much further back; for in each variation lies hidden *the mystery of a beginning*. We cannot tell why offspring should be like unto parent; how then should we know why it should sometimes be different?

So then Darwinism has real causes at its foundation, viz., the fact of variation and the inevitable operation of natural selection, determining the survival only of the fittest forms for the time and place. It is therefore a good hypothesis, so far. But is it a sufficient and a complete hypothesis? Does it furnish scientific explanation of (*i.e.*, assign natural causes for) the rise of living forms from low to high, from simple to complex, from protoplasm to simple plant and animal, from fish to flesh, from lower animal to higher animal, from brute to man? Does it scientifically account for the formation of any organ, show that under given conditions sensitive eye-spot, initial hand or brain, or even a different hue or texture, must then and there be developed as the consequence of assignable conditions? Does it explain how and why so much, or any, sensitiveness, faculty of response by movement, perception, consciousness, intellect, is correlated with such and such an organism? I answer, Not at all! The hypothesis does none of these things. For my own part I can hardly conceive that any one should think that natural selection scientifically accounts for these phenomena.

Let us here discriminate. To account scientifically for phenomena, or for complex series of phenomena, by assigning real and sufficient natural causes, is one thing. To believe that the phenomena have occurred in the course of nature, and have natural causal connection, is another. It is not natural selection which has led Mr. Darwin and many others to believe that life was "originally breathed by the Creator into a few forms or into one," and "that the production and extinction of the past and present inhabitants of the world has been due to secondary causes;" but it is the observed fact of likenesses and that of gradation from form to form which suggested the idea of an actual evolution from form to form having somehow taken place. Variation and natural selection are now assigned as causes or reasons of the evolution. Variation originates all the differences. Natural selection, determining which forms shall survive, reduces their number and intensifies their character. But Darwin may likewise consistently speak of his favorite principle as *a* cause of the evolution, it being that in the absence of which the evolution could not take effect. A *cause of variation* it certainly is not, but it is a necessary *occasion* of it, or of its progress. Because without natural selection to pave the way, the wheels of variation would at once be clogged and all progress be arrested. Variation provides that upon which natural selection operates; the operation of natural selection makes room for further variation, gives opportunity for variability to change its fashions and display its novelties; and so the two go on, hand in hand. But, although thus conjoined, there is always this difference between the two, that natural selection works externally, with known natural agencies, and in the light of common day; variation works internally, in darkness, and its agencies and ways are recondite and past finding out. Or, when we find out something, — as we may hope to do, — we only resolve a before unexplained phenomenon into two factors, one of them a now ascertained natural process, the other a something which still eludes our search. But we suppose it to be natural, although as yet unknown. Surely we are not to suppose that natural agencies cease just where we fail to make them out.

To proceed: what Darwinism maintains is that variation, which is the origination of small differences, and species-production, which represents somewhat larger differences, and genus-production, which represents still greater differences, are parts of a series and differ only in degree, and therefore have common natural causes whatever these may be; and that natural selection gives a clear conception of a way in which continually or occasionally arising small differences may be added up into large sums in the course of time. This is a legitimate and on the whole a good working hypothesis. The questionable point is whether the sum of the differences can be obtained from the individually small variations by simple addition. I very much doubt it. I doubt especially if simple addition is capable of congruously adding up such different denominations. That is, while I see how variations of a given organ or structure can be led on to great modification, I cannot conceive how non-existent organs come thus to be, how wholly new parts are initiated, how any thing can be led on which is not there to be taken hold of. Nor am I at all helped in this respect by being shown that the new organs are developed little by little.

The doubt is not whether the organs and forms were actually evolved in the course of Nature. I agree with Darwin that they probably were, and if so then doubtless under natural selection. And I cannot help thinking that Darwin would agree with me that the principle of natural selection does not account for it. That is, we both account for it all, only by assuming as an inexplicable fact that variation does occur to the whole extent of the extreme differences. All appears to have come to pass in the course of Nature, and therefore under second causes; but what these are, or how connected and interfused with first cause, we know not now, perhaps shall never know.

Now views like these, when formulated by religious instead of scientific thought, make more of Divine providence and fore-ordination than of Divine intervention; but perhaps they are not the less theistical on that account. Nor are they incompatible with "special creative act," unless natural process generally is incompatible with it, — which no theist can allow. No Christian theist can eliminate the idea of Divine intervention any more than he can that of Divine ordination; neither, on the other hand, can he agree that what science removes from the supernatural to the natural is lost to theism. But, the business of science is with the course of Nature, not with interruptions of it, which must rest on their own special evidence. Still more, it is the business of science to question searchingly all seeming interruptions of it, and its privilege, to refer events and phenomena not at the first but in the last resort to Divine will.

Moreover, "special creative act" is not excluded by evolutionists on scientific ground, is not excluded at all on principle, except by those who adopt a philosophy which antecedently rules out all possibility of it. Darwin postulates one creative act and a probability of more, and so in principle is at one with Wallace and with Dana, who insist on more.

But it has been said, and indeed is said over and over, even by thoughtful men, that, although Darwinism is not necessarily atheistic, yet, when once started it dispenses with further need of God. "Given [it is said] the laws which we find, then there is no more use for God, and all things have come out as we find them with none of his supervision. There may have been — we do not know — a God once; but law and not God, is the great Creator." A few words should dispose of this. First, by what right is it assumed that the Darwinian differs from the orthodox conception of law? In the next place, this line of argument applies equally to a series of creative acts separated by intervals, during which it could with the same reason (or unreason) be said that there is no use for God, that there may have been a God at times! So it cuts away the ground from under the Christian evolution which the writer quoted from allows, as well as from that which he deprecates. And it equally dispenses with use for God in Nature for the several thousand years which have passed since creation under the biblical view was finished, and the Creator "rested from all the work which he had made." There is no more validity in the argument in the one case than in the others.

A word or two upon the subject of creative acts occurring in time may not be out of place. These, when spoken of in the present connection, do not usually refer to the making of a new form of plant or animal *instanter* out of the dust of the ground. However it might have been when there was only one act of creation to think of, the enormous crudeness of such a conception when applied to a long succession of animals would now be seriously felt by every one. It is a phrase most used by those who accept the idea of the evolution of one species from another, but who feel the utter incompetence of known natural causes to account for it. In the absence of such causes, they, being theists, naturally (and I cannot say unphilosophically) assign the simpler and seemingly easier part of evolution to recondite natural causes which they are unable to specify, the more difficult or inscrutable to a diviner and more direct or supernatural act, which they liken to creation. I suppose they do not feel the necessity, as they have not the ability, to draw any definite line between what they think mere Nature may accomplish, and what they believe she cannot. Probably what they have in mind is mediate creation and not miracle. Perhaps they are convinced that if they could behold the birth of a species, they would see nothing more miraculous than in the birth of an individual. They mean that the springs of Nature are somehow touched by a new form or instance of force directed to some new end. Yet so they must be in a degree in the origination of a new race or variety. This whole conception of mediate creation is logically carried out to its extreme by my philosophical colleague, Professor Bowen, when he concludes that "not only every new species but that each individual living organism, originated in a special act of creation." *

So the difference between pure Darwinism

---

* North American Review for November, 1879, p. 463.

and a more theistically expressed evolution is not so great as it seemed. Both agree in the opinion that species are evolved from species, and that evolution somehow occurs in the course of Nature. Darwinism opines that the whole is a natural result of general causes such as we know of and in a degree understand, such as we recognize under the concrete terms of variability, heredity, and the like, — terms which we can estimate and limit only by reference to what we see coming to pass, — along with complex physical interactions which are more measurable and predictable. The very much that it has not accounted for by these causes and processes, it assumes may be in time accounted for by them, or by as yet unrecognized general causes like them. The specially theistic evolution referred to judges that these general causes cannot account for the whole work, and that the unknown causes are of a more special character and higher order. I think it does not declare that these are not secondary causes, and whether they would be ranked as natural causes would depend upon the sense in which the term Nature was at the moment used. Probably such evolutionists, if they had to give form to their conceptions, would vary in all degrees between the direct interposition of a supernatural hand at certain stages or crises, and that extreme extension of the Supernatural into and through the Natural which Professor Bowen reaches in the assertion that each individual living organism, as well as every new species, originated in a special act of creation. This, the complete assimilation of specific to individual origination, is simply Darwinism, expressed in less appropriate language. What the one calls "special act" the other, along with the rest of mankind, calls general process. The common principle of the Divine ordination of Nature, which the philosopher here asserts in a paradoxical way, the Darwinian implies, or even postulates, on appropriate occasions. The Darwinian *Naturalist*, I mean, not the monistic and agnostic philosopher, — from whom, so far, we have kept as clear as has Mr. Darwin in every volume and every line.

Suppose now that we are shut up to Nature for the evolution of the forms of living things. As theists, we are not debarred from the supposition of supernatural origination, mediate or immediate. But suppose the facts suggest and inferentially warrant the conclusion that the course of natural history has been along an unbroken line; that — account for it or not — the origination of the kinds of plants and animals comes to stand on the same footing as the rest of Nature. As this is the complete outcome of Darwinian evolution, it has to be met and considered.

The inquiry, what attitude should we, Christian theists, present to this form of scientific belief, should not be a difficult one to answer In my opinion, we should not denounce it as atheistical, or as practical atheism, or as absurd. Although, from the nature of the case, this conception can never be demonstrated, it can be believed, and is coming to be largely believed; and it falls in very well with doctrine said to have been taught by philosophers and saints, by Leibnitz and Malebranche, Thomas Aquinas, and Augustine. So it may possibly even share in the commendation bestowed by the Pope, in a recent sensible if not infallible allocution, upon the teaching of "the Angelic Doctor," and make a part of that genuine philosophy which the Pope declares to stand in no real opposition to religious truth. Seriously it would be rash and wrong for *us* to declare that this conception is opposed to theism. *Our* idea of Nature is that of an ordered and fixed system of forms and means working to ultimate ends. If this is our idea of inorganic nature, shall we abandon or depreciate it when we pass from mere things to organisms, to creatures which are themselves both means and ends? Surely it would be suicidal to do so. We may, and indeed we do, question gravely whether all this work *is* committed to Nature; but we all agree that much is so done, far more than was formerly thought possible; we cannot pretend to draw the line between what may be and what may not be so done, or what is and what is not so done; and so it is not for us to object to the further extension of the principle on sufficient evidence.

I trust it is not necessary to press this consideration, though it is needful to present it, in order to warn Christian theists from the folly of playing into their adversary's hand, as is too often done.

But I am aware that we have not yet reached the root of the difficulty. *We* are convinced theists. We bring our theism to the interpretation of Nature, and Nature responds like an echo to our thought. Not always unequivocally: broken, confused, and even contradictory sounds are sometimes given back to us; yet as we listen to and ponder them, they mainly harmonize with our inner idea, and give us reasonable assurance that the God of our religion is the author of Nature. But what of those — you will say — who are not already convinced of His existence? We thought that we had an independent demonstration of His existence, and that we could go out into the highways of unbelief and "compel them to come in;" that "the invisible things of Him from the creation of the world were clearly seen, being understood by the things that are made," "so that they are without excuse." We could shut them up to the strict alternative of Divinity or Chance, with the odds incalculably against Chance. But now Darwinism has given them an excuse and placed us on the defensive. Now we have as much as we can do, and some think more, to reshape the argument in such wise as to harmonize our ineradicable belief in design with the fundamental scientific belief of continuity in nature, now extended to organic as well as inorganic forms, to living beings as well as inanimate things. The field which we took to be thickly sown with design seems, under the light of Darwinism, to yield only a crop of accidents. Where we thought to reap the golden grain, we find only tares.

The outlook is certainly serious, yet not altogether disheartening. Perhaps we cannot now safely separate the wheat from the tares, but must let them grow together unto the harvest. Nobody expects in this world to ascertain the limits between design and contingency. Nobody expects to demonstrate any design, except his own to himself by consciousness; he cannot really prove his own to his bosom friend; though his assertion may give his friend, and his actions may give his enemy, convincing reasons for inferring it. But we are sure that every intellectual being has designs, that the reach and pervasiveness of design must be in proportion to the wisdom; and that the designs of the Author of Nature, if any there be, must be all-pervading and fathomless. Yet if they be wrought into a system of adaptations, some of the adaptations themselves may be such as irresistibly to suggest their reason to our minds. At least they suggest *reason*, even if we fail to apprehend, or wrongly apprehend, *the reason*. The sense that there is *reason why* is as innate in man, as that there is *cause whereby*.

Now, to adopt the apt words of Francis Newman,* "after stripping off all that goes beyond

* In Contemporary Review, 1878, p. 445, &c.

the mark of sober and cautious thought, there remain in this world fitnesses innumerable on the largest and the smallest scale, in which alike common sense and 'uncommon sense see design, and the only mode of evading this belief is by carrying out the cumbrous Epicurean argument to a length of which Epicurus could not dream. We cannot prove, we are told, that the eye was intended to see, or the hand to grasp, or the fingers to work delicately. Of course we cannot. But what is the alternative? To believe that it came about by blind chance. No science has any calculus or apparatus to decide between the two theories. Common sense, not science, has to decide, and the most accomplished physical student has in the decision no advantage whatever over a simple but thoughtful man."

Arrangements innumerable, extending through all nature, subserving all ends, of course involve innumerable contingencies. The theist is not expected to have any definite idea of the respective limits of these. He can only guess at the limits of intention and contingency in the actions of his nearest neighbor. The non-theist gains nothing by eliminating instances, unless he can eliminate all design from the system.

Until he does this, he gains nothing by showing that particular fitnesses come to pass little by little, and under natural causes. He cannot point to a time where there were no fitnesses, apparent or latent, and if he argues that all fitnesses were germinal in the nebulous matter of our solar system, he does not harm our case. The throwing of design ever so far back in time does not harm it, nor deprive it of its ever-present and ever-efficient character. For, as has been acutely said, "If design has once operated *in rerum natura* (as in the production of a first life-germ), how can it stop operating and undesigned formation succeed it? It cannot, and intention in Nature having once existed, the test of the amount of that intention is not the commencement but the end, not the first low organism, but the climax and consummation of the whole."\*

I am not going to re-argue an old thesis of my own that Darwinism does not weaken the substantial ground of the argument, as between theism and non-theism, for design in Nature.†

---

\* Mozley, Essays, ii. 412. See also Lord Blachford in The Nineteenth Century, June, 1879, p. 1035.

† Darwiniana: Essays and Reviews pertaining to Darwinism. New York, D. Appleton & Co., 1876.

I think it brought in no new difficulty, though it brought old ones into prominence. It must be reasonably clear to all who have taken pains to understand the matter that the true issue as regards design is not between Darwinism and direct Creationism, but between design and fortuity, between any intention or intellectual cause and no intention nor predicable first cause. It is really narrowed down to this, and on this line all maintainers of the affirmative may present an unbroken front. The holding of this line secures all; the weakening of it in the attempted defence of unessential and now untenable outposts endangers all.

I have only to add a few observations and exhortations addressed to Christian theists.

If intention must pervade every theistic system of Nature, if we give credit to Mr. Darwin when in this regard he likens his divergence from the orthodox view to the difference between general and particular Providence, is it safe to declare that his theory, and his denial that particular forms were specially created, are practically atheistical? I might complain of this as unfair: it is more to my purpose to complain of it as suicidal. It is in effect holding a theistic conception of Nature for our private use, but acting on the opposite when we would discredit an unwelcome theory. Or else it is trusting so little to our own belief that we abandon it as soon as any weight is laid upon it. As soon as you do this, by conceding that the evolution of forms under natural laws militates against design in Nature, you are at the mercy of those reasoners, who, looking at the probabilities of the case from their own point of view, coolly remark that:—

"On the whole, therefore, we seem entitled to conclude that, during such time as we have evidence of, no intelligence or volition has been concerned in events happening within the range of the solar system, except that of animals living on the planets."\*

You may say that implicit belief of intention in Nature affords an insufficient foundation for theism. But you are not asked to ground your theism upon it, nor upon the whole world of external phenomena.

You may reiterate that you cannot believe that all these events have occurred under natural laws. Nothing hinders your assuming what you need from the supernatural; but

---

\* Clifford, Sunday Lectures, quoted in The Spectator.

allow that the need of other minds may not be identical with yours.

As I have said before, what you want is, not a system which may be adjusted to theism, nor even one which finds its most reasonable interpretation in theism, but one which theism only can account for. That, it seems to me, you have. An excellent judge, a gifted adept in physical science and exact reasoning, the late Clerk-Maxwell, is reported to have said, not long before he left the world, that he had scrutinized all the agnostic hypotheses he knew of, and found that they one and all needed a God to make them workable.

When you ask for more than this, namely, for that which will compel belief in a personal Divine Being, you ask for that which He has not been pleased to provide. Experience proves that the opposite hypothesis is possible. Some rest in it, but few I think on scientific grounds. The affirmative hypothesis gives us a workable conception of how "the world of forms and means" is related to "the world of worths and ends." The negative hypothesis gives no mental or ethical satisfaction whatever. Like the theory of the immediate creation of forms, it explains nothing.

You inquire, whither are we to look for independent evidence of mind and will "concerned in natural events happening within the range of the solar system." Certainly not to the court of pure physical science. For that has ruled this case out of its jurisdiction by assuming a fixed dependence of consequent upon antecedent throughout its domain. There are plenty of phenomena to which it cannot assign known causal antecedents; but it supplies their place at once, either by assuming that there is a physical antecedent still unguessed, or by inventing one in an hypothesis. It deals in effects and causes, and knows nothing of ends. It has no verdict to render against our case, for it does not entertain it, and has no jurisdiction under which to try it. But its wiser judges do not insist that theirs is the only court in the realm.

We have not to go beyond Nature for a jurisdiction, which may be likened to that of Equity, since it enforces specific performance, and which adds to causes and effects the consideration of ends. Biology takes cognizance of the former, like physics, of which it is on one side a part, but also of ends; and here ends (which mean intention) become a legitimate scientific study. The natural history of ends becomes consistent and reasonably intelligible under the light of evolution. As the forms and kinds rise gradually out of that which was well-nigh formless into a consummate form, so do biological ends rise and assert themselves in increasing distinctness, variety, and dignity. Vegetables and animals have paved the earth with intentions. The study and the estimate of these is quite the same, under whatever view of the mode in which the structures and beings that exemplify them came to be.

The highest of these exemplars is himself conscious of ends. He pronounces that critical monosyllable *I*. I am, I will, I accomplish ends. I modify the outcome of Nature. Here, at length, is something "on the planets" which "has been concerned in events;" and in my opinion it is just now a good and useful theistic view which connects this something with all the lower psychological phenomena that preceded and accompany it. Our wills, in their limited degree, modify the course of Nature, subservient though that be to fixed laws. By our will we make these laws subserve our ends. We momently violate the uniformity of Nature. But we do not violate the *law* of the uniformity of Nature. Is it not legitimate, is it not inevitable, that a being who knows that he is a will, and a power, and a successful contriver, should explain what he sees around and above him by the hypothesis of a higher and supreme will? A will which has disposed things in view of ends in establishing Nature, and which may, if need be, dispose to particular and timed ends, either with or without perceptible suspension of the law of the uniformity of Nature.

The question I ask has been adversely answered, substantially as follows: It may be that in the first instance men can hardly avoid predicating a being who has done and is doing all this. Nevertheless a trained mind soon reaches the incongruity of it, at least " as concerns any events which have happened within the range of the solar system." For the belief that a supernatural power has so acted contradicts that very belief in the uniformity of Nature upon which all scientific reasoning and practical judgments rest.

To this it is well rejoined, that the ultimate scientific belief on which our reason reposes "is *that* belief in the uniformity of Nature which is equivalent to a belief in the law of universal causation; which again is equivalent to a belief that similar antecedents are always followed by similar consequents. But this belief is in no way inconsistent with a belief in supernatural interference." * If the principle of the uniformity of Nature asserted that every natural effect is, and has ever been, preceded by natural causes, *then* it would be in terms inconsistent with supernatural interference and with supernatural origination of the system. But science does not give us nor find any such principle. All scientific beliefs " are in themselves as true and as fully proved if supernatural interference be possible as they are if such interference be impossible. A law does no more than state that under certain circumstances (positive and negative) certain phenomena will occur. If on some occasions these circumstances, owing to supernatural interference, do not occur, the fact that the phenomena do not follow proves nothing as to the truth or falsehood of the law." * If such interference violates the law of the uniformity of Nature, the human will, and all wills, and all direction of material forces to ends, are every day violating it.

It is also urged that giving particular direc-

* Balfour (Arthur). A Defence of Philosophic Doubt, p. 329. The note on the Discrepancy between Religion and Science is particularly pertinent.

tion in a special act would be an addition to the *plenum* of force in the universe, and therefore a contradiction to the recently acquired scientific principle of the conservation of energy. The answer may be this. It is not at all certain that all direction given to force expends force; it is certain that, under collocations, a minute use of force (as pulling a hair-trigger or jostling a valve) may bring about immense results; and, finally, increments of force by Divine action in time, of the kind in question, if such there be, could never in the least be known to science.

The only remaining supposition that I now think of is the crude one that thought and will are functions of the body, secretions as it were of the organ through which they are manifested, "psychical modes of motion." Then, as has well been said, they must be correlated with physical modes of motion, at least in conception; but it is conceded by all sensible thinkers that thought cannot be translated into extension, nor extension into thought. Now, since the only conceivable source of physical force is supernatural power, still more must this be the only conceivable source of thought.

There is an old objection which threatens to

undermine the ground on which we infer Divine will from the analogy of human; namely, that our wills, being a part of the course of Nature and amenable to its laws, their movements, though seemingly free, are as fixed as physical sequences. Upon this insoluble problem we have nothing practical to say, except to admit that so much of choice is determined by antecedent conditions and the surroundings, by hereditary bias, by what has been made for the individual and inwrought into his nature, that, granting the will has an element of freedom, it may be in effect a small factor. I can only urge that it is not an *insignificant* factor. As to this, a pertinent although homely suggestion came to me in the remark of a humble but shrewd neighbor, to the effect that he found the difference between people and people he dealt with was really very little, but that what there is was very important. So facts and reasonings may shut us up to the conclusion that the will, sovereign as it seems to the user, is practically a small factor in the determination of events. But what there is makes all the difference in the world in man!

And now, as to man himself in relation to evolution. I have no time left for the discussion of questions which naturally interest you more than any other, but which, even with time at disposal, are not easy to treat. I will not undertake to consider what your attitude should be upon a matter which connects itself with grave ulterior considerations; but I will very briefly and frankly intimate what views I think a scientific man, religiously disposed, is likely to entertain.

To pursue the illustration just ventured upon: The anatomical and physiological difference between man and the higher brutes is not great from a natural-history point of view, compared with the difference between these and lower grades of animals; but we may justly say that what corporeal difference there is is extremely important. The series of considerations which suggest evolution up to man, suggest man's evolution also. We may, indeed, fall back upon Mr. Darwin's declaration, in a case germane to this, that "analogy may be a deceitful guide." Yet here it is the only guide we have. If the alternative be the immediate origination out of nothing, or out of the soil, of the human form with all its actual marks, there can be no doubt which side a scientific man will take. Mediate creation, derivative origination will at once be accepted; and the mooted question comes to be narrowed down to this: Can the corporeal differences between man and the rest of the animal kingdom be accounted for by known natural causes, or must they be attributed to unknown causes? And shall we assume these unknown causes to be natural or supernatural? As to the first question, you are aware, from my whole line of thought and argument, that I know no natural process for the transformation of a brute mammal into a man. But I am equally at a loss as respects the processes through which any one species, any one variety, gives birth to another. Yet I do not presume to limit Nature by my small knowledge of its laws and powers. I know that a part of these still occult processes are in the every-day course of Nature; I am persuaded that it is so through the animal kingdom generally; I cannot deny it as respects the highest members of that kingdom. I allow, however, that the superlative importance of comparatively small corporeal differences in this comsummate case may justify any one in regarding it as exceptional. In most respects, man is an exceptional creature. If, however, I decline to regard man's origin as exceptional in the sense of directly supernatural, you will understand that it is because, under my thoroughly theistic conception of Nature, and my belief in mediate creation, I am at a loss to know what I should mean by the exception. I do not allow myself to believe that immediate creation would make man's origin more divine. And I do not approve either the divinity or the science of those who are prompt to invoke the supernatural to cover our ignorance of natural causes, and equally so to discard its aid whenever natural causes are found sufficient.*

It is probable that the idea of mediate creation would be more readily received, except for a prevalent misconception upon a point of genealogy. When the naturalist is asked, what and whence the origin of man, he can only answer in the words of Quatrefages and Virchow, " We do not know at all." We have traces of his existence up to and even anterior to the latest marked climatic change in our temperate zone : but he was then perfected man; and no vestige of an earlier form is known. The believer in direct or special creation is entitled to the advantage which this negative evidence gives. A totally unknown ancestry has the characteristics of nobility. The evolutionist

---

\* See Baden Powell, On The Order of Nature, p. 163.

can give one satisfactory assurance. As the wolf in the fable was captious in his complaint that the lamb below had muddied the brook he was drinking from, so those are mistaken who suppose that the simian race can have defiled the stream along which evolution traces human descent. Sober evolutionists do not suppose that man has descended from monkeys. The stream must have branched too early for that. The resemblances, which are the same in fact under any theory, are supposed to denote collateral relationship.

The psychological differences between man and the higher brute animals you do not expect me now to discuss. Here, too, we may say that, although gradations abridge the wide interval, the transcendent character of the superadded must count for more than a host of lower similarities and identities; for, surely, what difference there is between the man and the animal in this respect is supremely important.

If we cannot reasonably solve the problems even of inorganic nature without assuming initial causation, and if we assume for that supreme intelligence, shall we not more freely assume it, and with all the directness the case may require, in the field where intelligence at length develops intelligences? But while, on the one hand, we rise in thought into the supernatural, on the other we need not forget that one of the three old orthodox opinions, — the one held to be tenable if not directly favored by Augustine, and most accordant to his theology, as it is to observation, — is that souls as well as lives are propagated in the order of Nature. Here we may note, in passing, that since the "theologians are as much puzzled to form a satisfactory conception of the origin of each individual soul as naturalists are to conceive of the origin of species," and since the Darwinian and the theologian (at least the Traducian) take similar courses to find a way out of their difficulties, they might have a little more sympathy for each other. The high Calvinist and the Darwinian have a goodly number of points in common.*

View these high matters as you will, the outcome, as concerns us, of the vast and partly comprehensible system, which under one aspect we call Nature, and under another Providence,

---

* See an article on Some Analogies between Calvinism and Darwinism, by Rev. G. F. Wright, in the Bibliotheca Sacra, January, 1880.

and in part under another, Creation, is seen in the emergence of a free and self-determining personality, which, being capable of conceiving it, may hope for immortality.

"May hope for immortality." You ask for the reasons of this hope upon these lines of thought. I suppose that they are the same as your own, so far as natural reasons go. A being who has the faculty — however bestowed — of reflective, abstract thought superadded to all lower psychical faculties, is thereby *per saltum* immeasurably exalted. This, and only this, brings with it language and all that comes from that wonderful instrument; it carries the germs of all invention and all improvement, all that man does and may do in his rule over Nature and his power of ideally soaring above it. So we may well deem this a special gift, the gift beyond recall, in which all hope is enshrined. None of us have any scientific or philosophical explanation to offer as to *how* it came to be added to what we share with the brutes that perish; but it puts man into another world than theirs, both here, and — with the aid of some evolutionary ideas, we may add — hereafter.

Let us consider. It must be that the Eternal can alone impart the gift of eternal life. But He alone originates life. Now what of that life which reaches so near to ours, yet misses it so completely? The perplexity this question raises was as great as it is now before evolution was ever heard of; it has been turned into something much more trying than perplexity by the assurance with which monistic evolutionists press their answer to the question; but a better line of evolutionary doctrine may do something toward disposing of it. It will not do to say that thought carries the implication of immortality. For our humble companions have the elements of that, or of simple ratiocination, and the power of reproducing conceptions in memory, and — what is even more to the present purpose — in dreams. Once admit this to imply immortality and you will be obliged to make soul coextensive with life, as some have done, thereby well-nigh crushing the whole doctrine of immortality with the load laid upon it. At least this is poising the ponderous pyramid on its apex, and the apex on a logical fallacy. For the entire conception that the highest brute animals may be endowed with an immortal principle is a reflection from the conception of such a principle in ourselves; and so the farther down you carry it, the wider and more egregious the circle you are reasoning in.

Still, with all life goes duality. There is the matter, and there is the life, and we cannot get one out of the other, unless you define matter as something which works to ends. As all agree that reflective thought cannot be translated into terms of extension (matter and motion), nor the converse, so as truly it cannot be translated into terms of sensation and perception, of desire and affection, of even the feeblest vital response to external impressions, of simplest life. The duality runs through the whole. You cannot reasonably give over any part of the field to the monist, and retain the rest.

Now see how evolution may help you; — in its conception that, while all the lower serves its purpose for the time being, and is a stage toward better and higher, the lower sooner or later perish, the higher, the consummate, survive. The soul in its bodily tenement is the final outcome of Nature. May it not well be that the perfected soul alone survives the final struggle of life, and indeed "then chiefly lives," — because in it all worths and ends inhere; because it only is worth immortality, because it alone carries in itself the promise and potentiality of eternal life! Certainly in it only is the potentiality of religion, or that which aspires to immortality.

Here I should close; but, in justice to myself and to you, a word must still be added. You rightly will say that, although theism is at the foundation of religion, the foundation is of small practical value without the superstructure. Your supreme interest is Christianity; and you ask me if I maintain that the doctrine of evolution is compatible with this. I am bound to do so. Yet I have left myself no time in which to vindicate my claim; which I should wish to do most earnestly, yet very deferentially, considering where and to whom I speak. Here we reverse positions: you are the professional experts; I am the unskilled inquirer.

I accept Christianity on its own evidence, which I am not here to specify or to justify; and I am yet to learn how physical or any other science conflicts with it any more than it conflicts with simple theism. I take it that religion is based on the idea of a Divine Mind revealing himself to intelligent creatures for moral ends.

We shall perhaps agree that the revelation on which our religion is based is an example of evolution; that it has been developed by degrees and in stages, much of it in connection with second causes and human actions; and that the current of revelation has been mingled with the course of events. I suppose that the Old Testament carried the earlier revelation and the germs of Christianity, as the apostles carried the treasures of the gospel, in earthen vessels. I trust it is reverent, I am confident it is safe and wise, to consider that revelation in its essence concerns things moral and spiritual; and that the knowledge of God's character and will which has descended from the fountain-head in the earlier ages has come down to us, through annalists and prophets and psalmists, in a mingled stream, more or less tinged or rendered turbid by the earthly channels through which it has worn its way. The stream brings down precious gold, and so may be called a golden stream; but the water — the vehicle of transportation — is not gold. Moreover the analogy of our inquiry into design in Nature may teach us that we may be unable always accurately to sift out the gold from the earthy sediment.

But, however we may differ in regard to the earlier stages of religious development, we shall agree in this, that revelation culminated, and for us most essentially consists, in the advent of a Divine Person, who, being made man, manifested the Divine Nature in union with the human; and that this manifestation constitutes Christianity.

Having accepted the doctrine of the incarnation, itself the crowning miracle, attendant miracles are not obstacles to belief. Their primary use must have been for those who witnessed them; and we may allow that the record of a miracle cannot have the convincing force of the miracle itself. But the very reasons on which scientific men reject miracles for the carrying on of Nature may operate in favor of miracles to attest an incoming of the supernatural for moral ends. At least they have nothing to declare against them.

If now you ask me, What are the essential contents of that Christianity which is in my view as compatible with my evolutionary conceptions as with former scientific beliefs, it may suffice to answer that they are briefly summed up in the early creeds of the Christian Church, reasonably interpreted. The creeds to be taken into account are only two,— one commonly called the Apostles', the other the Nicene. The latter and larger is remarkable for its complete avoidance of conflict with physical science. The language in which its users "look for the resurrection of the dead" bears — and doubtless at its adoption had in the minds of at least some of the council — a worthier interpretation than that naturally suggested by the short western creed, namely, the crude notion of the revivification of the human body, against which St. Paul earnestly protested.

Moreover, as brethren uniting in a common worship, we may honorably, edifyingly, and wisely use that which we should not have formulated, but may on due occasion qualify,— statements, for instance, dogmatically pronouncing upon the essential nature of the Supreme Being (of which nothing can be known and nothing is revealed), instead of the Divine manifestation. We may add more to our confession: we all of us draw more from the exhaustless revelation of Christ in the gospels; but this should suffice for the profession of Christianity. If you ask, must we require that, I reply that I am merely stating what I accept. Whoever else will accept Him who is himself the substance of Christianity, let him do it in his own way.

In conclusion, we students of natural science and of theology have very similar tasks. Nature is a complex, of which the human race through investigation is learning more and more the meaning and the uses. The Scriptures are a complex, an accumulation of a long series of records, which are to be well understood only by investigation. It cannot be that in all these years we have learned nothing new of their meaning and uses to us, and have nothing still to learn. Nor can it be that we are not free to use what we learn in one line of study to limit, correct, or remodel the ideas which we obtain from another.

Gentlemen of the Theological School, about to become ministers of the gospel, receive this discourse with full allowance for the different point of view from which we survey the field. If I, in my solicitude to attract scientific men to religion, be thought to have minimized the divergence of certain scientific from religious beliefs, I pray that you on the other hand will never needlessly exaggerate them; for that may be more harmful. I am persuaded that you, in your day, will enjoy the comfort of a much better understanding between the scientific and the religious mind than has prevailed. Yet without doubt a full share of intellectual and traditional difficulties will fall to your lot. Discreetly to deal with them, as well for yourselves as for those who may look to you for guidance, rightly to present sensible and sound doctrine both to the learned and the ignorant, the lowly and the lofty-minded, the simple believer and the astute speculatist, you will need all the knowledge and judgment you can acquire from science and philosophy, and all the superior wisdom your supplications may draw from the Infinite Source of knowledge, wisdom, and grace.

www.ingramcontent.com/pod-product-compliance
Lightning Source LLC
Chambersburg PA
CBHW021945160426
43195CB00011B/1224